Bygones presented by Dick Joice
on Anglia Television

BARNS AND GRANARIES
IN NORFOLK

Sheridan Ebbage

The Boydell Press · Ipswich

Published by The Boydell Press Ltd
PO Box 24 Ipswich IP1 1JJ

ISBN 0 85115 079 9

Printed in Great Britain by
Lowe & Brydone Printers Limited, Thetford, Norfolk

FOREWORD

You don't have to be in the company of Norfolk born Sheri Ebbage for long to understand why after being at school she spurned all the things more conventional girls thrive on, like special 18th birthday parties – but shot off to Canada – not the easy way but roughing it – log cabins and all – then back and a quick dash round the art centres of Europe – Paris, Rome, Venice, Florence etc. Finally turning up in my office at Anglia House, demanding to know what I knew about Norfolk Barns. There wasn't much I could tell her that she didn't know already, except that I thought it was time that someone did something about recording them, for they were fast disappearing as the use for them dwindled in the light of modern agricultural needs. She said "I'm going to take Norfolk Barns as the subject for my thesis for a diploma". Promising to look at it when it was finished, I wished her the best of luck and off she went.

About two years later back she came with a beautifully prepared work . . . photographs taken by herself and a B.A. degree. It's an understatement to say her work deserves to be published, but I certainly have great pleasure in recommending it to anyone who appreciates these beautiful buildings that so typify agriculture and Norfolk builders at their best.

DICK JOICE

INTRODUCTION

For over 600 years the barn dominated the farmstead, by its sheer size. It was the heart of the farm: the farming year began and ended in the barn.

Every type of farm building has its own history: the barn is perhaps the most interesting for it has a special significance on the farm. Harvest marked the culmination of the year's work: the produce of harvest was stored in the barn. A good harvest meant a barn packed high with corn: that meant a full stomach. Thus the degree of success of the farming year could be measured by the contents of the barn.

People were proud of their barns, and gave them as much architectural detail as they could afford.

The abundance of barns in Norfolk is due to the fact that it is primarily a corn-growing region. Arthur Young, in the early 19th century, called it "the greatest barley county in the kingdom". William Marshall thought the barns of Norfolk "superior to those of every other county: numerous and spacious".

Throughout its long history, there was little change in the design of the barn. Its function, apart from the storing of unthreshed corn, was to provide a suitable floor for the threshing and winnowing. The advent of the threshing machine in the late 18th century, rendered these processes obsolete. The barn was unable to adapt to this and became redundant.

Because of the lack of a good building stone in Norfolk, the barns are built of a wide variety of local materials: the farmers used those which were nearest to hand. The distribution of walling materials relates loosely to the surface geology of the area.

Each of these materials provides a different architectural effect: thus each area has its own local flavour. A flint barn with a thatch roof in North Norfolk, has quite a different character from a brick barn with a pantile roof in Central Norfolk. The individual character of the barn is enhanced by the details such as the gable-ends, the 'owl-holes' and the ventilation 'loop-holes'.

The existence, the size, and the distribution of barns, undoubtedly reflects the national and local farming patterns in history. The enormous barns of North-West Norfolk stored corn for huge estates: these light soil regions were enclosed as large fields in the 18th century, by wealthy landowners. These barns differ considerably from those of South Norfolk, in the 'wood-pasture' area, a region by-passed by the great 18th century rebuilding. This is an area of heavy clay, which was enclosed at an early date, and still today, remains a region of smallholdings.

The barn has no future. In isolated cases one might be considered worth preserving and maintaining, or possibly converting for residential use. Many are used for the storage of implements or bagged grain. Many more stand redundant and rotting.

The barn no longer dominates the farmstead even in its size, for the huge modern farm buildings dwarf even the largest. This is not sad: it is a natural development. The barn was a functional building: without that function it is simply a shell. Those barns of architectural value must be conserved, but let others make way for the natural development of farming.

THE PRINCIPLE OF THE BARN

The word 'barn' is derived from the Old English 'bern', meaning a place for storing grain. The barn housed the unthreshed corn after harvest, and was the building in which the threshing took place. It is generally an oblong building, most commonly divided into 3 areas: this type of barn is known as a '3-stead' barn. The middle area, known as the 'middlestead', was used as a threshing floor, and has a pair of doors on each of the barn walls, through which the waggons could enter. Either side of the middlestead are the 'sidesteads' or 'goafsteads' where the unthreshed corn was stored.

A natural development from the 3-stead barn were the 5- or 7-stead barns. A 5-stead barn would have 4 pairs of doors and 3 goafsteads: a 7-stead barn would have 6 pairs of doors and 4 goafsteads. The size of the barn was dependent on the amount of corn to be stored and threshed and thus reflected the size of the farm, and the wealth of its owner. (See Illustration 1.) The smallest barn I have yet seen was only 33 feet long (Washingford Farm, Bergh Apton); whilst the longest was about 184 feet long (Hales Court, near Loddon).

Tithe barns were instituted for the purpose of storing a tenth of each local farmer's produce. This was claimed by the church: it was a law intended for the support of the clergy, and continued up to the end of the 18th century. G. M. Trevelyan wrote in his *English Social History*: "The tenth sucking pig went to the parson's table: the tenth sheaf was carried off to his tithe barn." This lucrative system enabled the tithe barns to be much grander in their construction. They are usually twice the size of an average farm barn. Henry Warren calls them "cathedrals of labour" and likens the Paston barn to "the same noble proportions as the nave of a great church".

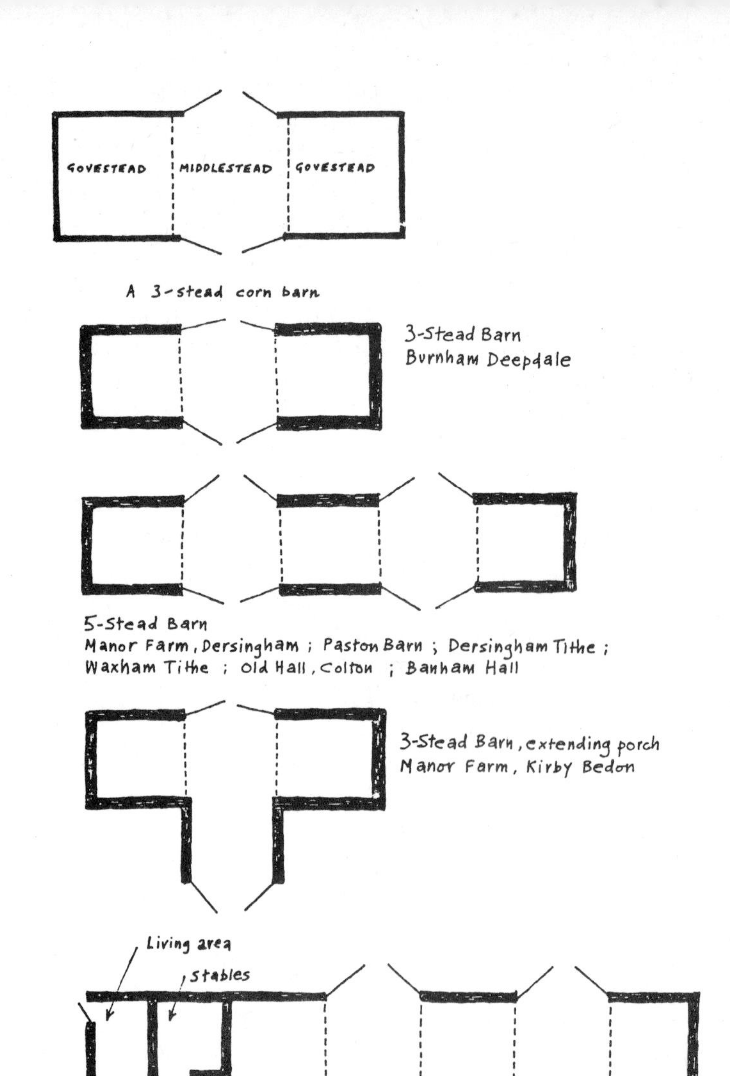

1. *Types of barns.*

Grange barns are also much larger than the average 3-stead barns. These were built and used by the monasteries and priories. Until the dissolution of the monasteries in 1536–9, 650 of them were estimated to own a quarter of the total agricultural land in England. They practised agriculture on an enormous scale, and were largely self-supporting. These buildings were originally repositories for grain, but after the dissolution of the monasteries, the term was more loosely applied to tithe barns, or barns in monastic establishments. Castle Acre Priory Grange was measured as 160 feet long just before it fell down in 1838.

Barns were often built to provide living accommodation: the barn at Hales Court has living accommodation at the east end, with a fireplace and window. This is next to the stables, which were also enclosed in the barn: the living accommodation was probably provided for the man who looked after the horses. (See Illustration 1.) Above this area is a loft.

Field barns are located some way from the farmstead, and were used for the storage of corn, hay, implements, or to provide accommodation for young cattle or sheep. The sides are enclosed, and they are similar, though generally smaller in design, to a 3-stead barn. These barns sometimes provided living accommodation, so that the cowman could stay with his herd at calving time. Field barns are generally located on larger farms, especially where the weather is more severe. There are more in the north-west regions of Norfolk than in any other.

The corn was harvested and either bound into 'sheaves' (pronounced 'shoofs') or carted loose in the waggons, and taken into the barn. The loaded waggon was brought into the farmyard and through the big, main doors into the barn. These would be swung open 180°, flat against the wall. The width of the doors was governed by the width of the fully laden waggon and a man walking alongside: I found this to be usually 12 feet. The height of the front doors was the height of the wall. This was governed by the height of a fully laden waggon.

The waggon was drawn in, and stopped on the middlestead, while the men off-loaded the corn into the goafsteads. Ewart Evans quotes an age-old custom in his book *The Farm and the Village*:

> "One practise connected with this stage of the harvest was called 'riding the goaf' in Suffolk. As the loose barley was unloaded on to the goafstead, a boy rode a quiet old farmhorse round and round on the corn, tramping it down. The main purpose was to pack as much corn as possible into the bay: but in treading the corn in this way much of it was shaken out of the ear, and thus made a start on the threshing . . . The main problem came when horse and rider could rise no higher and had to be got down. One man told me that they left a rough

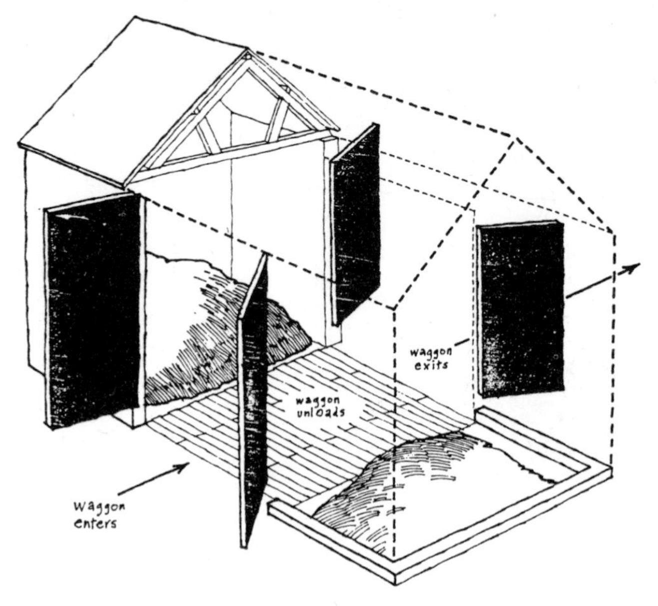

2. *The 3-stead barn showing the passage of the waggon.*

sort of ramp at one side of the corn, and the horse slid down this onto the middlestead. Another described how a strong rope was thrown over one of the barn's tie-beams and then fixed to the horse's harness. The men eased him down by taking much of his weight on the rope".

Robert Bloomfield's autobiographical poem *The Farmer's Boy*, written at the end of the 18th century, describes his experiences of 'riding the goaf':

> . . . e'en humble Giles
> Who joys his trivial services to yield:
> Amidst the fragrance of the open field:
> Oft doom'd in suffocating heat to bear
> The cobweb'd barn's impure and dusty air;
> To ride in murky state the panting steed
> Destin'd aloft th' unloaded grain to tread,
> Where, in his path as heaps and heaps are thrown,
> He rears and plunges the loose mountain down;
> Laborious task! with what delight when done
> Both horse and rider greet the unclouded sun".

This accomplished, the horse and empty waggon would be taken out straight through the smaller pair of doors at the back of the barn, and off to collect some more sheaves. (See Illustration 2.)

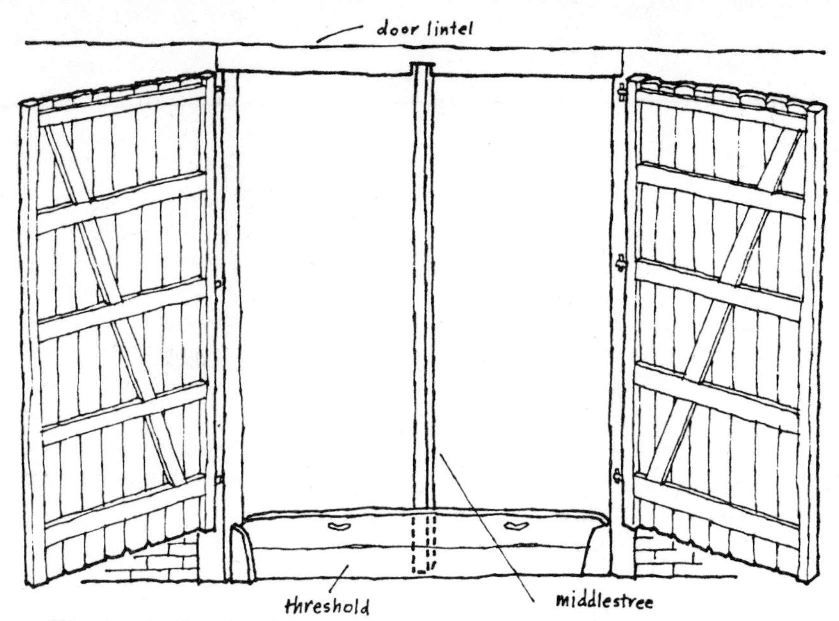

3. *The threshold and middlestree.*

With harvest over the barn would be sealed up until the winter months, when the threshing would begin. The doors on barns were built so they could be swung open 180°, and hooked back against the wall. When they were closed, they were held in position by a piece of wood known as a 'middlestree'. This was a length of wood, square in section, which fitted into a hole in the door lintel at the top, and a hole in the ground below. The 'threshold' was then lifted into place, slotting into the width of the doorway along the ground. This was 2 or 3 planks of wood, joined together to fit the width of the door. Two handles can usually be seen on the threshold: these are to lift it in or out. The threshold fitted in front of the middlestree, and stands generally, about 1′ 6″–2′ high. The doors lock onto the middlestree, overlapping the threshold slightly. At harvest time the middlestree and threshold were taken out to allow the waggons to enter. Most, but not all barns, have a middlestree and threshold. (See Illustration 3.) A barn at Banhall Hall, Banham, has a hole cut out of the threshold, about 5″ in diameter. The owner told me that this was a 'cat-hole', which allowed the farm cats to enter the barn and dispose of the vermin. I imagine that it must have let the vermin in as well!

A much more common feature on a barn, is the 'owl-hole'. This can be seen in a variety of shapes and sizes, about 1 foot square being the maximum. The 'owl-hole' is usually on the gable end wall just under the ridge. This allowed owls to enter the barn and dispose of the vermin.

Although 'sealed' from the weather, a barn was never sealed from rats. About December or later, the threshing would begin.

Ewart Evans writes:

> "The flail, the tool they used for this purpose, has always been as much a symbol of the corn harvest as the sickle itself: and its history is almost as ancient . . . The flail succeeded the beating-stick, and improved on its action. For the beating stick, as it was in one piece, had to be used in a straight up-and-down action to beat out the corn. But the flail can be used in a circular action like a whip, which enabled its operator to keep up a steady rhythm. In fact, the word flail derives from the Latin 'flagellum', a whip. The implement is basically two sticks joined by a flexible knot . . . The stick or handle is usually made of ash, and is twice as long as the swingel, the part that strikes the straw to shake out the grain".

The threshing was done in the middlestead on an elm or poplar board floor: the unthreshed corn was brought from the goafsteads. It was boring, arduous work, carried out by 5 or 6 men at a time. Ewart Evans describes the men as keeping up certain rhythms like bell-ringers, to break up the monotony. The wooden-floor helped these subtle patterns, and allowed them to be heard some way off.

After threshing was completed, the 'winnowing' or 'dressing' of the corn, would begin: the dust and chaff that had got mixed up with the grain had to be cleaned out before it could be weighed and sent to the mill. The doors were opened, and the grain thrown into the draught that was created. This blew away the lighter bits of dust and chaff, whilst the heavier grain fell to the ground.

The double doors on barns are sometimes divided into 4 sections: the bottom pair stand about 5 feet high from the threshold; the top pair, normally about 8–12 feet high. The bottom doors could be opened to allow a man to enter, without opening the whole door. If it were a sufficiently windy day, they would be opened for the winnowing.

The grain was cast into the draught using special wooden shovels called 'casting shovels'. They were made of wood because this was less likely to bruise the grain. Wooden sieves with fine iron-wire or cane meshes were also used for this purpose. The dust would fall through, whilst the bits of straw could be picked out.

Ewart Evans describes a 19th century winnowing machine called a 'blower':

> "an arrangement of 5 fans fixed to a central spindle that was turned by a crank. One operator turned the crank while another fed the corn into the hopper. A third carried away the dressed

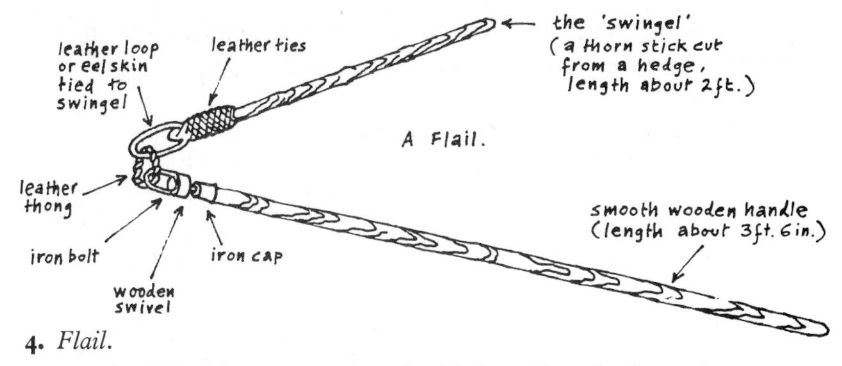

4. *Flail.*

grain. The blower was placed with its tail to the barn-door so that the wind helped blow away the chaff."

After dressing, the grain was lifted up with short-handled hoes, and scooped up with 'maunds', large wooden containers, and into 'bushels' – the standard measure. This comprised of a circular wooden drum with metal bands and iron handles. The top of the bushel was levelled off – with great exactness – with a 'strike', or stick. If, perchance, the grain settled too much, or the strike was drawn across the top wrongly, the bushel had to be emptied out and the process started again.

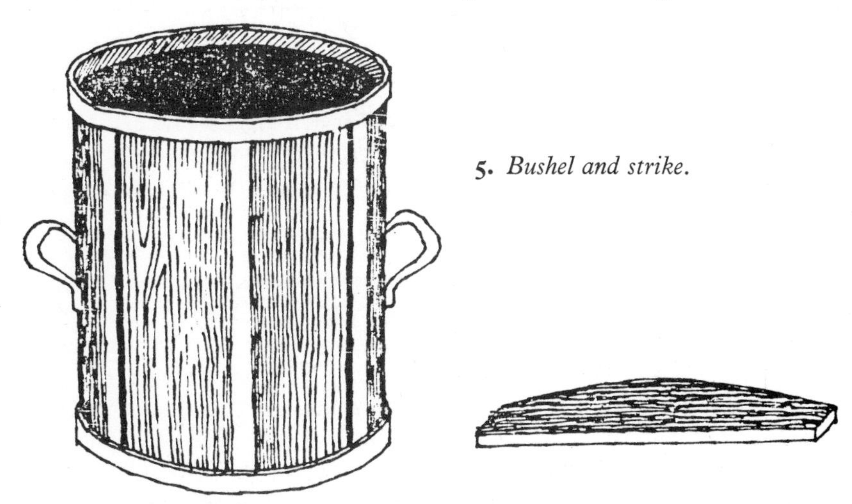

5. *Bushel and strike.*

The grain was generally taken straight to the mills after dressing. Sometimes a farmer would take a sample of his corn to the market or corn hall, before he had begun threshing, sometimes after. The chaff from the corn would be stored and used as fodder for cattle and horses. It was often mixed with chopped turnips or mangolds. The straw was stacked and used for litter, thatching or fodder.

The barn remained empty for a period of 8–9 months, until the next harvest.

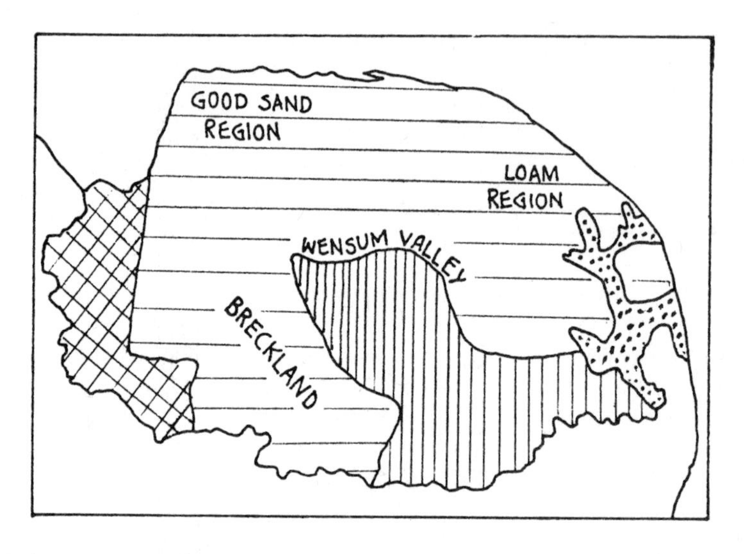

MAP SHOWING THE 'WOOD-PASTURE' AREA AND THE 'SHEEP-CORN' REGION IN NORFOLK IN THE 16TH AND 17TH CENTURIES.

- The 'wood-pasture' area
- The 'sheep-corn' region
- The Fens
- The Broads

Reproduced from 'The Sheep-Corn Husbandry of Norfolk in the 16th and 17th centuries' by Keith Allinson

6. *The 'wood-pasture' area in Norfolk.*

THE BARN IN AGRICULTURAL HISTORY

The oldest barn still standing in Norfolk is at Hales Court: it was built at the end of the 15th century. It marks the change from the old manorial system of farming. A new class of prosperous landowners and yeoman farmers was emerging. They bought up the old manorial farms and the peasant holdings, amalgamated them, and leased them to tenant farmers. The monastic and ecclesiastical establishments which had once accounted for vast areas of farmland, collapsed with the dissolution of the monasteries in 1536–9. Their lands were dispersed and bought up by lay landowners. The villages of Norfolk had one lord, as opposed to many, by 1650.

The population increased from under 3 million people in 1500, to approximately 4 million in 1600. Food production was increased by adding to agricultural area, rather than by improving existing farming methods. The reclamation of the fens, the seacoasts, marshes and ancient

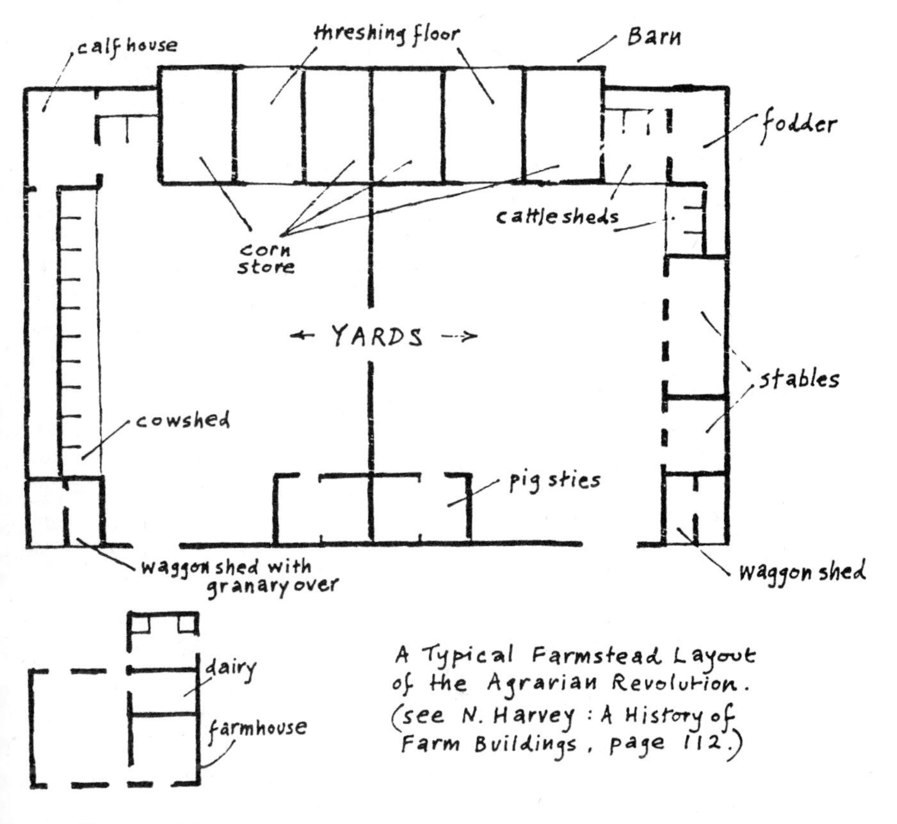

7. *Farmstead layout.*

forests was under way creating new farmland. Only 1 % of the total land area of Norfolk was actually enclosed in 1517, however, when a Commission of Inquiry was set up.

Norfolk appears to have been divided into two main areas during the 16th and 17th centuries: the 'wood-pasture' area, and the 'sheep/corn' region. (See Illustration 6.) The 'wood-pasture' area was the region of heavy clay, covering most of South Norfolk. Some corn was undoubtedly grown here, but farmers concentrated on cattle. The farms in this area were enclosed at an earlier date than those in the 'sheep/corn' region. The fields were smaller, and to this day it remains an area of small-holdings.

The second area, the 'sheep/corn' region was an area of good, light soil: this was where the best corn was grown. The region was characterised by the 'foldcourse' system. This worked on the principle that the heaths and commons were used as summer pastures for the sheep, whilst the winter pastures were the strips of arable land, laid fallow after harvest. This resulted in extensive dunging of the light soils, and enabled farmers to produce heavier cereal crops such as wheat and barley. This is the area in which the great 18th century rebuilding was carried out, under wealthy landowners. The lands were mostly enclosed at a later date; the fields and farms are much larger than those in the 'wood-pasture' area. Coke of Holkham, and 'Turnip' Townsend practised their new agricultural methods in this area in the 18th century. I found the barns in this area, to be larger on average, than other regions of Norfolk.

The function of the barn remained unchanged throughout the 16th and 17th centuries. The fundamental change in agriculture in the 16th century, was the gradual transition from subsistence farming, to commercially based agriculture. One of the reasons for this was the increase in urban population. The new class of yeoman farmers created a demand for improved domestic architecture. This affected the barn and other farm buildings, only in a general improvement of building methods. The farmer was still dependent on local materials and local skill.

Nigel Harvey writes in his *History of Farm Buildings*:

> "Economically, patterns of regional specialisation were developing to meet the new commercial opportunities, but the days of the specialist farms were still far in the future. There was, therefore, little change in the demands of the farm on the farmstead and developments were few and simple . . . There were no fundamental changes in this period in the inherited alliance of barn, beasthouse and yard."

The barn continued to dominate the farmstead. The medieval

tradition of positioning the farm buildings at the end of the house had long died out. These buildings were now separated from the farmhouse, though certain functions such as the dairy processes and beer making continued in the farmhouse.

The Dutch and French immigrants who settled in Norfolk in the 16th and 17th centuries were skilled craftsmen. They helped improve the quality of building, as well as popularising the use of brick as a building material. The barns at the Old Hall, Colton, dated 1666, and Manor Farm, Kirby Bedon, dated 1693, have the 'crow-step' and 'Dutch' gables popular at the time. (See Illustrations 13 and 14.)

Although they were largely a phenomenon of the Middle Ages, tithe barns continued to be built until the 17th century: the Dersingham tithe barn, for instance, is dated 1671. Architecturally tithe barns are generally more impressive than other barns. But they cannot be considered representative of English farming and the average size of the grain crops at the time.

Harvey writes:

> "The large barns described as 'tithe' barns are among the glories . . . of the English countryside. But they survived because they were exceptional. . . . Indeed many of the huge and gracious barns of the clergymen were hardly farm buildings at all. Some, presumably fulfilled the functions of the conventional farm barn in which the harvest of the surrounding fields was threshed and stored. But others were the central depots of huge agricultural estates, housing the harvests of scattered farms run by bailiffs and any rents which were paid in corn. The true descendant of the tithe barn is not the farmer's grainstore but the merchant's warehouse. The tithe barn, therefore, illustrated ecclesiastical wealth".

The tithe system must have been lucrative to necessitate such large barns: for they are generally twice the size of the average farm barns. The system allowed for building on a grander, more decorative scale. The alternating hammer-beam roof of the tithe barn at Paston, built in 1581, is reminiscent of a church. The Dersingham tithe barn, with its variety of materials and decorative gable ends is another example.

By the beginning of the 1700s, there were approximately 5½ million people in Britain. Attitudes towards agriculture were becoming more enlightened. Enclosure was still piecemeal however: from 1700–10, there was only one Enclosure Act passed in Parliament. By the end of the 17th century, the Fens had been almost completely reclaimed, along with the marshes and forests. Owing to the lack of further land to reclaim, improvements had to be made in agricultural methods.

The old 'foldcourse' system in West Norfolk had broken down. The pioneer work in crop rotation, which was to solve this problem, was being carried out by Sir Richard Weston. Later, 'Turnip' Townshend developed the 'Norfolk' four-course shift.

By 1750, 38 Enclosure Acts had been passed in Parliament. The lands of West Norfolk now came under extensive enclosure. The new 'whole-year' lands excluded sheep, and Norfolk farmers in the 'sheep/corn' region, began to devote themselves to the cultivation of corn.

The new farms created by the enclosures, needed new farm buildings. The old farmsteads needed extension and improvement to provide for the increasing corn crops. In this great rebuilding, a basic pattern of farmstead design began to emerge. The position of the farm buildings in relation to their function and the needs of the farmer, became the main design criteria. By-products of the barn, such as straw, went to the cattle-houses and stables for use as litter or fodder.

Harvey describes a typical farmstead layout of this period:

> "They . . . took the form of a series of buildings round open yards, sometimes in a square, but more usually on three sides of a square. . . . The yards faced south to catch the sun and avoid some at least of the rain-bringing south-west winds. They were sheltered on the north by the most substantial building on the farm, the barn – on large farms two or three barns, ideally one for each type of grain. . . . Into these barns from the rickyard to the north of them came corn for threshing and winnowing, on the central floor, out of them came grain. . . . From here, too, came straw for littering the various livestock buildings, and, above all, the yards. From this north range and at right-angles to it ran the wings which formed the yards and contained a medley of buildings . . . arranged in accordance with this need for straw . . . convenient to the cattleyards, sometimes in a barn, sometimes in stacks, stood the haystore".

Isolated field barns, used for storing grain or implements, or housing cattle, began to be built in the late 18th century.

In 1776, Thomas William Coke inherited the family estate at Holkham. The rent roll then brought in £2,200 a year. By 1816 Coke had increased this to £20,000 a year. Coke was considered to be one of the greatest agricultural innovators of the Agrarian Revolution. He advocated long leases to tenants, strict rotation of crops, new techniques of drainage, and improved stock breeding. One of the many improvements he made was to replace the old farm buildings with magnificent

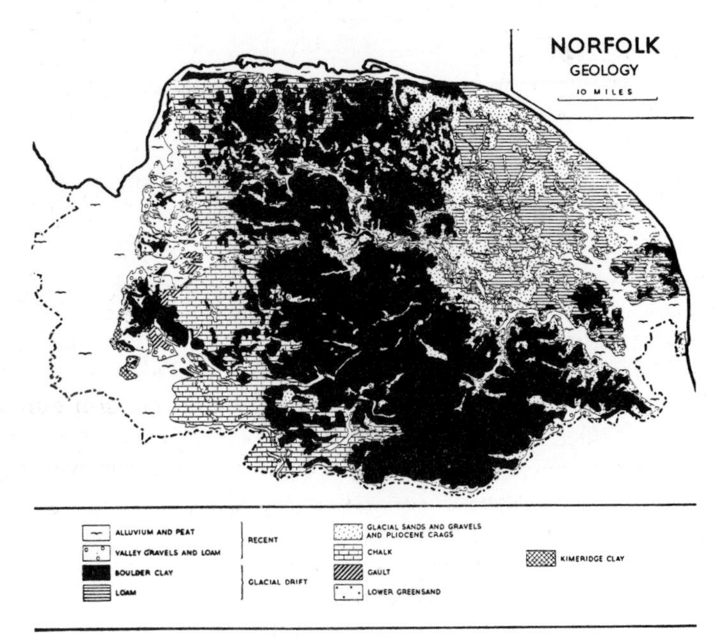

Fig. 22. Norfolk: Surface geology.
Based on Geological Survey Quarter-Inch Sheets 12 and 16.

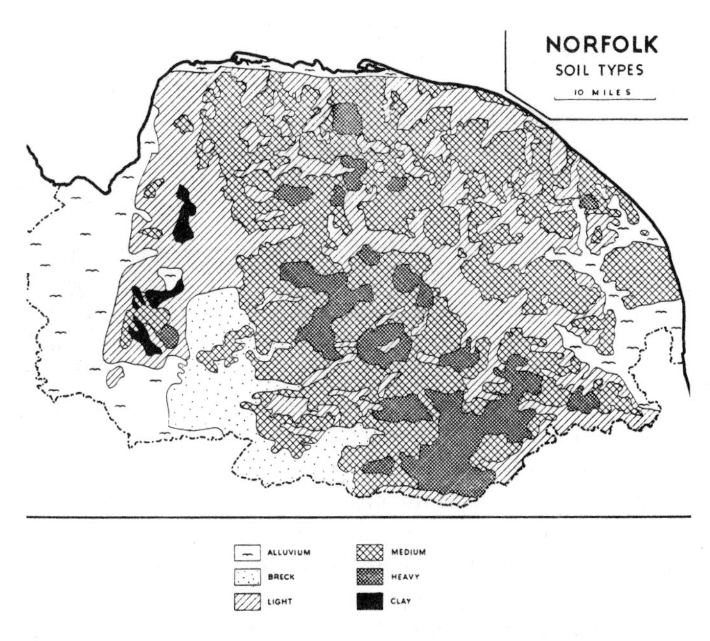

8. *The surface geology of Norfolk.*

9. *The Great Barn, Holkham.*

new ones. These were to be a demonstration of the new principles of the 'model' farm. The 'Great Barn' at Holkham, built 1790–2, was thought by some to be "the first barn in England" – "a magnificent structure… a striking example of the increasing importance of farm buildings in this period". The enormous, 5-stead barn is entirely enclosed by sheds, stables and storehouses. It is completely symmetrical in design. (See Illustration 9.)

The life of the corn barn was to be short lived, however, with the invention of the threshing machine. In 1786, Andrew Meikle, a Scots millwright, built a mechanical threshing machine. The existing barns could accommodate these machines quite easily.

The high roofs, designed for the swinging flails became unnecessary. The vast storage space was no longer needed for the machines could cope with a continuous flow of sheaves. The 'stack-yard' system evolved, in which the unthreshed corn was stacked outside until threshing time. This changed the position of the barn in the farmyard once again, since the most efficient layout was now to have the barn end on between stackyard and cattleyard: in the past, the barn had faced the stackyard on one side and the cattleyard on the other.

In 1794, Nathaniel Kent made his survey of Norfolk agriculture, and wrote:

> "Farmers are very averse to stacking: though wheat is preserved better and sweeter on staddles than in a barn: they are always crying out for barn room: and they are certainly indulged in a greater proportion of it than farmers in any other county.... There are many single barns which have lately been erected.... This is certainly wrong for such buildings make a great waste of timber, and are unnecessary, and moreover very bad examples, as one farmer will always covet a similar thing to what he sees his equal in possession of . . .".

Within twenty years the threshing machine was widely adopted. It reduced corn wastage, put an end to the dependence on unreliable farm labour, and enabled the farmer to market his crop earlier, as the machine could thresh the entire crop in half the time.

The machines were housed in existing barns. Arthur Young, the agricultural writer, commented on Coke's "very large machine" at Holkham:

> " . . . it occupies too great a space in one of the finest barns in England. It prevents the storing of near 300 quarters of barley".

Young prophesied that the mechanical thresher "promised speedily to put an end to all barn building". Although his prophecy was essentially correct, the building of barns continued up to the latter half of the 19th century, but on a much smaller scale. The barn at Binham, dated 1860 (see Illustration 20) and the barn at East Raynham, dated 1870 (see Illustration 21) are much smaller than the barns built previous to this time.

From 1800–10, the last Enclosure Acts were passed in Parliament, bringing the total up to 3,036. The first official census in 1801, gave the population of England and Wales as 9 millions: by 1811 this had increased to 10,164,256. This rapid increase forced farmers to grow much larger corn crops. The Napoleonic Wars were a lucrative time for farmers. After this period, however, prices dropped and there was large-scale unemployment. The Corn Laws were first passed in Parliament in 1815, causing further hardship to farmers. The New Corn Law was introduced in 1828, resulting in the Anti-Corn Law League of 1839. A series of bad harvests before 1842, and a severe trade depression beginning in 1839 aggravated the situation. This provided little incentive to farmers to construct new farm buildings.

The steam engine was another radical change which rendered the barn defunct. A portable steam-engine threshing machine was developed, which was taken to the stacks in the fields, and saved the farmer the task of moving the stacks to the farmyard. The slow process of threshing was converted to a 2 or 3 day operation. The flood of cheap North American

10. *Barn at Leicester Square Farm, Syderstone.*

grain in the 1870s did little to increase the farmer's wealth. It is interesting that barn building seems to have virtually ceased after this decade. By this time the barn had become a "historic relic", but because it was usually substantially built, it usually survived rather than being pulled down. There were undoubtedly some farms which still threshed the corn in the barn using the flail, but this was probably the exception rather than the rule.

With the advent of the combine harvester in the late 1920s, the barn, neglected for the previous half-century, became functional once more – even if not performing the task for which it was originally intended. The combine harvester could cut and thresh the corn in a single operation. It still left the problem of drying however, which was the purpose of the stacks. They were an aerated store which allowed the corn to dry out naturally, so that it did not overheat or gather moulds. Thus the grain-dryer evolved, and the barn was often used to house it.

This is how the barn is sometimes found in use today. The floors have generally been concreted and the doors and ventilation holes blocked in. Buildings designed to house grain dryers first appeared in the 1930s. On large arable farms, the barns could usually be adapted: the barn at Manor Farm, Dersingham, was one of the first barns to be used in this way.

Barns are very uneconomic buildings to maintain and insure however. They would be insured for their present replacement cost, which, considering the enormous area, and size of the walls and roofs of some barns, would be a considerable amount of money. Many thatched barns have had their roofs replaced with asbestos sheeting: this no doubt decreases the insurance premium, as it lessens the fire risks. Owing to the rising cost of building materials many farmers feel that their barns are no longer worth maintaining.

MATERIALS

In the construction of his barn, a farmer would use the materials nearest to hand. A farmer in the north of Norfolk, where flint is readily available, would have been most unlikely to transport clay from the southern end of the county to make a wattle and daub barn! His choice then, was governed by the most easily obtainable materials, and depending on his wealth, the cheapest.

R. W. Brunskill says in his *Handbook of Vernacular Architecture*:

> "Broadly speaking there were two structural systems available . . . mass construction and frame construction. In the former, the loads of roof and floors were carried to foundations by means of walls which also provided the weather protective envelope to the activities carried on within the building. In the latter, the loads of roof and floors were carried by a frame which concentrated these loads until they were redistributed by the foundation; the weather protective envelope was non-loadbearing . . . and could be independent of the frame".

Geologically, the foundation of the county is chalk. Much of this is buried beneath a foundation of boulder clays, sands and gravel. Flint is found in great quantities in the glacial sands, gravels and crags, and also in river-deposits. Owing to the lack of indigenous building stone, flint has acquired great importance as a material in the county. The wide belt of chalky boulder clay which runs through Norfolk, forms a natural source of brickearth. There were an enormous amount of kilns: Faden's map of 1797 shows a wide distribution. Clay for pantiles was often dug from the same quarries as the bricks. Carstone is found in the Lower Greensand, and can be seen in the Kings Lynn – Sandringham – Hunstanton strip. (See Illustration 8.)

BRICK

Brick is probably the most extensively used building material in Norfolk barns: it is certainly the most widely distributed. It is a strong, adaptable, fire-resisting material, and can be used alone without reinforcement. Owing to their regular size and shape, bricks are used for door and window jambs, and on the corners, of barns built of irregular materials such as flint and chalk. Different colour bricks are often used to produce diaper and chequerboard patterns. The Flemish and Dutch immigrants who settled in Norfolk promoted the use of brick, and in particular, the

11. *Barn at Hales Court*

decorative use of brick. One can see the characteristic 'crow-step' and 'Dutch' gables throughout Norfolk.

Bricks vary enormously in colour: the barn at Hales Court is a warm poppy-red, whilst Coke's Great Barn, at Holkham, is a yellow-grey. Like all the landowners Coke had his own kiln near Holkham: the bricks made here were considered to be amongst the finest in England. All the farm buildings on Coke's estate were built of brick. Two of the most impressive barns in Norfolk are his Great Barn at Holkham and the barn at Leicester Square Farm, Syderstone. (See Illustrations 9 and 10.) Both are completely symmetrical in design. The Great Barn, built 1790–2, is a 5-stead barn, with a Neo-Classical extended porch over each pair of doors. Sheds and stables are built around the barn, according to the principles of the 'model farm'. It has a slate roof: there is a ridge of raised 'headers' running underneath the eaves.

The barn at Leicester Square Farm, built in 1791, is of red brick, and has 3 sets of doors each with a slightly extended porch. It has a pantile roof, and like the Great Barn, some decorative brickwork. (See Illustration 10.)

The brick barn at Hales Court is possibly the longest in Norfolk, measuring some 184 feet. It was built by James Hobart, the Attorney-General to Henry VII, as part of Hales Court, at the end of the 15th century. It has 13 crow-step gables at each end, and 3 storeys of ventilation loops on the long walls. Between the loops on the front wall are the lozenge-shaped brick patterns popular on Tudor buildings. On the back wall of the barn are diagonal patterns in the brick. The moat which encircled Hales Court once ran round the back of the barn. When this

12. *Barn at Hales Court*

dried out, the 3 pairs of doors on the back wall were put in. Living accommodation with a fireplace was provided at the east end of the barn, hence the chimney stack on the east end wall. (See Illustrations 11 and 12.)

The barn at the Old Hall, Colton, has a fine set of crow-step gables. This barn is built of red brick, except for a few black bricks placed at random. The date of the barn, 1666, is on the front of the barn in framed letters. There is some interesting brick decoration around the ventilation loops, under the eaves, and on the consoles at the corners of the barn. (See Illustration 13.)

A fine set of 'Dutch' gables can be seen on the barn and granary at Manor Farm, Kirby Bedon. This is a small 3-stead barn with an extended porch. It is built of red brick, with flint footings and a pantile roof. The wrought-iron numbers on the end walls date it as 1693. (See Illustration 14.)

The brick barn at Bury's Hall, Carlton Rode, has a round owl-hole at the apex of the gable ends. The barn also has two sets of diagonal ventilation loops on each of the long walls. (See Illustration 44.)

Although brick is more usually seen as mass-walling, it is sometimes found used as infilling on timber-framed buildings: this is known as 'brick-nogging'. A most unusual barn of this ilk can be seen at the Old Hall, East Tuddenham. The house is Jacobean, and I imagine the barn is contemporary. The end walls of the barn have post-and-truss timber frames: the timbers running vertically and horizontally. The long walls

13. *Barn at The Old Hall, Colton*

of the barn have timber posts. The footings are brick. The brick nogging runs horizontally, not diagonally. (See Illustration 15.)

Dove House Barn at Morston, built of flint and brick in 1674, has brick jambs and corners. The date of the barn is picked out in brick on the front wall. (See Illustration 19.)

FLINT

Flint is probably the second most widely used building material in Norfolk barns. It was the cheapest material since it could be picked off the fields. When it was found in chalk quarries it was regarded as throw-away material, since it was chalk that people were seeking.

Flints can be used in two ways as a material: whole and split. When used whole, flints have a whitish appearance, as the 'rind' which covers the flint, is visible. They are laid in rough courses, and mortar is used as an infilling. When flints are split they give a flatter surface to the wall. The colour of the flint may then vary enormously from light grey to black, and the white rind is then only visible around the edge. To make up the thickness of a wall 1 yard wide, maybe 4 or 5 flints would be used.

Flint is a completely waterproof material. It is often used as a damp proof course, and for footings. The barn at Manor Farm, Kirby Bedon, has footings of split flints. (See Illustration 14.)

The tithe barn at Paston, built in 1581 by Sir William Paston, has walls of whole flints. Brick and stone are used on the door and loop jambs,

14. *Barn and Granary at Manor Farm, Kirby Bedon.*

quoins and buttresses. It is completely symmetrical in design. It is very narrow for its length, measuring 163 feet by 24 feet. The roof is 60 feet high at the apex, and is thus very steeply pitched. This adds to the feeling of the barn's enormous length. The roof is of reed thatch and there are raised gable ends. (See Illustrations 16 and 17.)

15. *Barn at The Old Hall, East Tuddenham.*

The tithe barn at Waxham is very similar in design to the Paston barn, in its roof construction. It is a 7-stead barn, as opposed to the Paston which is a 5-stead barn. The walls are of whole flints, and the quoins, jambs and buttresses are of stone and brick. On one end wall, there are diagonal patterns made with red brick in amongst the flint. This is a wider barn than Paston: it measures about 161 feet by 36 feet. The roof is of reed thatch. (See Illustration 18.)

Dove House Barn, Morston, built in 1674, has whole flint walls. This is a 3-stead barn, with long goafsteads. The roof is of pantiles. The overall length is 74 feet by 21 feet. (See Illustration 19.)

A small 3-stead barn at Binham, dated 1860, has cobble and brick walls. The cobbles are uncoursed and used as an exterior facing material: the interior walls are of whole flints. The quoins and jambs are of red brick. There are 3 narrow ventilation loops on the end walls, and an owl-hole by the apex. The roof is of pantiles. The barn measures 43 feet by 21 feet. (See Illustration 20.)

A completely symmetrical barn at East Raynham, built in 1870 by the Townshends, has brick and cobble walls. The quoins, jambs and footings are red brick, as is the ridge along the eaves underneath the roof. Cobble walls are more tightly packed than whole flint, owing to their more regular size and shape. The roof on this barn is 'hipped', and is covered with slate. (See Illustration 21.)

16. *Tithe barn, Paston.*

There is a small field barn near Letheringsett which has cobble walls. There is an extended porch on the front of the barn, with only a single door. The quoins and jambs are red brick, and it has a hipped pantile roof. The barn was built in 1851, according to the date above the door. (See Illustration 22.)

CARSTONE

Carstone "is a stone of coarse, pebbly or gritty consistency, strongly impregnated with iron oxide, and therefore always some shade of brown, ranging from 'café-au-lait' to deep chocolate." Norfolk people often call it 'gingerbread stone' on account of its colour. It becomes harder on exposure but nowhere can it be said to weather well, nor is it of good enough quality to invite fine workmanship. Yet apart from flint, it is the only building material in Norfolk. The principal quarries were at Snettisham.

The 'House Barn' at Manor Farm, Dersingham, is built entirely of carstone. The blocks are small, and irregular in size and shape. They are tightly packed together with a mortar infilling. An occasional black brick can be seen on the main walls. There are two sets of double doors with wooden lintels. The barn appears to have been constructed in two sections: i.e. one 3-stead barn was built onto the end of an existing 3-stead barn. There are diagonal ventilation patterns in the long walls.

17. *Tithe barn, Paston.*

The overall length is 72 feet by 24 feet. It has a pantile roof. The interior walls are a combination of brick and carstone. The quoins and jambs are brick, and there are 3 courses of brick beneath the eaves. (See Illustration 23.)

Two hundred yards from this barn, stands the Dersingham tithe barn. (See Illustration 24.) This 5-stead barn, dated 1671, is a combination of carstone, 'clunch' and brick. The footings, which rise up to 4 feet, are carstone, and there is a layer of carstone beneath the eaves, about 1 foot thick. Between the two carstone courses are large blocks of chalk. Red brick is used for the quoins, the jambs and the buttresses on the long

18. *Tithe barn, Waxham.*

19. *Dove House barn, Morston.*

walls. The end walls are built entirely of carstone, excepting two ridges of brick, and the ventilation jambs. Mortar is used as an infilling throughout. The 'crow-step' gables at each end of the barn are built of brick. The barn measures 98 feet by 27 feet, and has a pantile roof.

CHALK

The chalk found used as a building material on the barns in Norfolk, is usually called 'clunch'. It is a hard, marly variety of chalk. Although it can be used on its own, clunch is likely to weather better if used in combination with another material such as flint or brick. Although chalk blocks can be cut quite regular in size and shape, brick is used for the quoins and jambs.

The chalk blocks which make up the long walls of the Dersingham tithe barn are approximately 1 foot 6 inches long, by 9 inches deep. They have a rough surface, and are a very light grey.

There is a long 5-stead barn at Abbey Farm, Barton Bendish, which has chalk and brick walls. The barn appears to have been built in two sections. At one end the chalk blocks are laid in courses with mortar infilling. The other end of the barn has walls built of small, square chalk bricks placed at random with red bricks. A soft red brick is used on the quoins, jambs and on the large ventilation loops on the front wall. The end wall has carstone chips galletted into the mortar between the chalk blocks. (See Illustration 25.)

A similar barn can be seen at Manor Farm, South Creake. This barn has flint footings, brick quoins and jambs, and chalk walls. It is a long, 7-stead barn, and has a pantile roof. (See Illustration 31.)

A small 3-stead barn with brick and chalk chequerboard walls can be seen at Burnham Deepdale. The clunch bricks are laid in courses, alternating with red bricks. Brick is used on the quoins and jambs. The overall length is 48 feet by 20 feet. There is an owl-hole beneath the ridge on each of the end walls. The roof is of pantiles.

The Grange Barn at Castle Acre Priory, which has now completely disappeared, had a huge timber frame with chalk infilling. The blocks of chalk were then given a rough casing of flint, to make them weather resistant.

WEATHERBOARDING

Horizontal weatherboard is customary on Norfolk barns. The planks were pegged, and later nailed to vertical posts. Each board covered the one below with a single lap.

Carpenters would use whatever wood was to hand. Oak or elm were

20. *Barn at Binham.*

21. *Barn at East Raynham.*

preferred, but deforestation and subsequent rising prices forced them to use imported softwoods.

Weatherboard barns are usually tarred. As a building material it has few insulating qualities, and constitutes a considerable fire risk. It never lasted more than a century. It is probable that the weatherboard barn with a medieval frame, as it exists today, once had clay or brick infill: the timber frames would often shift and loosen the infill material.

There is a 5-stead barn at Banham Hall, with a timber frame covered with weatherboard. The pantile roof is steeply pitched, and the barn measures 90 feet by 24 feet. This is the barn with the 'cat-hole' in the threshold: it is a round hole 5″ in diameter, and 3″ off the ground. (See Illustration 27.)

The barn at Greys Farm, Great Witchingham, which was recently pulled down, was a weatherboard barn with a thatch roof. The end walls were of red brick. At one end was a stable and a loft above. The barn had brick footings up to a height of 3 feet. It was a 5-stead barn, and measured 77 feet by 20 feet. (See Illustration 28.)

Weatherboard was sometimes used on a gable end, as at Church Farm, Alderford: there is a dovecot on one of the gable ends. A similar gable and dovecot can be seen on the barn at Whissonsett Hall, dated 1778. (See Illustration 45.)

CLAY

Clay (the real name is 'daub'), has been used as a building material in Norfolk for at least 300 years. The clay is dug, spread out in a layer and the stones are removed. Short bits of straw are added, which act as a binding agent. It is then trampled by a horse. If the clay is to be made into 'lumps', it is then put into wooden moulds and left to dry out naturally in the sun. The dried blocks measure approximately 4″ × 18″ × 12″. Clay is sometimes used, puddled with straw, to build up a solid wall without moulds: this is called 'cob'. Another method is 'wattle and daub': hazel sways are woven together to form a 'wattle', and the clay is then pushed into this.

Clay is quite fire-resistant, and is a good insulator. It is a very cheap material in South Norfolk where it can often be simply dug out of a neighbouring field. Its major weakness is water: since the clay is not fired, it is still pervious and will crumble if in contact with water. Clay barns are usually tarred as a barrier against the rain.

The 3-stead barn at Manor Farm, Tacolneston, has a timber frame with clay infilling. A date carved in the tie-beam shows it to have been built in 1779. The clay infill is unmoulded, though the smaller pair of doors at the back of the barn have been replaced with clay lumps. The straw thatch roof has a steep pitch, a common characteristic of the barns of South Norfolk. The barn measures 72 feet by 30 feet. There are no ventilation loops.

There is a tiny 3-stead barn at Washingford Farm, Bergh Apton. It has untarred clay walls, and measures only 33 feet by 18 feet. The man who owns it thinks it was made from clay dug from the neighbouring field. He told me it used to be a field barn for a big, neighbouring farm,

22. *Field barn at Letheringsett.*

23. *House Barn, Manor Farm, Dersingham.*

although it is next to his house, which is also made from clay. He used to work at Hales Court, and can remember 'riding the goaf' as a lad, in the enormous barn. This barn has no timber frame, and is very crude in construction: it is impossible to date this sort of barn. If a bit fell off, the farmer would simply build it up again. It was a continual process of renewal. There are two very rough loop-holes on the front wall of the barn. The gable ends are weatherboard, and the roof is reed thatch. (See Illustration 29.)

I have noticed that the clay barns of South Norfolk tend to be smaller on average than the flint and brick barns of the North and West regions.

They are narrower, the roofs pitch steeply and are generally of thatch. Most of the clay and weatherboard barns in this area have extended porches. The heavy clay of South Norfolk forms a natural source of cheap building material.

TIMBER FRAMES

The majority of timber frame barns I found to be in the South of Norfolk, in the 'wood-pasture' area. This region had thicker forestation than the North and the West regions, and clay was readily available as an infilling material.

There was little wood in Norfolk: Alec Clifton Taylor writes in *The Pattern of English Building*:

> "In the Breckland area of Norfolk and Suffolk, a Royal proclamation in 1604 ordained that new houses must henceforth have their walls and window-frames of brick or stone. . . . Hence in these areas half-timbering as it decayed, was replaced with other materials, usually brick or flint, so that in some districts scarcely a single timber-framed building can be found. By the beginning of the 18th century oak was in short supply everywhere".

Any oak available for barn building was required for the huge tie-beams that spanned the roofs, being the strongest and most durable wood. Elm was used preferably for floors and weatherboarding, and served as an inferior substitute for oak when the latter became scarce. By the mid-16th century, soft-woods were imported: pine and spruce could be worked more quickly than the hard-woods, and were in this sense preferred. Soft-woods were used for rafters and suchlike, oak and elm continued to be used for loadbearing beams.

The timber framework of the grange barn at Castle Acre Priory had ten massive oak timbers, ranged five on each side. These formed the lateral walls of the barn. They were placed at 22 feet intervals and 8 feet from the exterior wall. This gave the barn the appearance of a church nave. The supporting columns joined over the middle space at a height of 22 feet, above which rose arches and spans of wood which joined at the summit, 33 feet from the ground. The barn was last recorded and drawn in May 1838. Blocks of chalk were used as an infilling, with a flint casing on the exterior wall.

The 3-stead barn at the Old Hall, East Tuddenham, has a box-frame on one end wall, with square panels. The front wall of the barn has vertical panels. Brick is used as an infilling, and the footings are also brick. The posts are set approximately 3 feet apart. (See Illustration 15.)

The vertical posts on the walls at Banham Hall are only 1 foot apart.

(See Illustration 27.) Those at Grey's Farm, Great Witchingham, were 2 feet apart.

The timber frames do not normally rise from the ground. Alec Clifton-Taylor writes:

> "First there is a base (the footings) of some solid material such as stone, flint or sometimes large balks of oak laid horizontally; then a wooden sill placed along the top of this base. Strong upright posts or 'studs' (the word derives from the Old English 'stuthu' meaning a pillar or post), often with the butt-ends uppermost, are mortised into the sill, whilst their upper-ends are tenoned into another horizontal beam called a summer or bresummer. . . ."

INTERIOR ROOF STRUCTURES

Roof framing has always been the carpenter's problem since neither stone nor brick could be used to bridge the walls.

The width of the barn was dependent on the available lengths of timber: this was usually 20–25 feet, but could be as much as 36 feet. The choice of roof covering affected the pitch of the roof: thatch, for instance, was laid at a pitch of not less than 60°, whilst pantiles were laid at a pitch of not more than 45°, and generally 30°–35°. The weight of the material governed 'the required dimensions of roof timbers'.

The roof structures, like timber frames were all prefabricated in the carpenter's yard.

H. L. Edlin writes in *Woodland Crafts in Britain*:

> "The carpenter selected his timbers in the woods, buying them as trees. He hewed them into shape whilst still green, and laid out the work flat so that each part would fit its neighbour; appropriate marks were cut into each piece so that it could be identified and its position determined. Tenons were carefully shaped, mortices cut with a narrow-bladed mortising axe, and dozens of oak pegs cleft to fit into auger holes bored to receive them".

The identification marks – usually roman numerals – are often visible in the tie-beams and braces. They are clearly visible on the timbers of the Paston barn. This roof is possibly the most impressive, and certainly the most quoted, of any barn in Norfolk. It spans an area of 163 feet by 24 feet, and has alternating hammer-beams and queen-post trusses. (See Illustrations 31 and 32.) There are 18 bays, and the roof is 60 feet high at the apex.

Similar to the Paston barn in its construction is the nearby tithe barn at Waxham. This barn has flint walls, with an alternating hammer-beam and queen-post truss roof, spanning an area of 162 feet by 36 feet. Both are thatched.

The roof of the brick barn at Hales Court has 11 tie-beams, 13 bays, and spans an area of 184 feet by 27 feet. The barn is divided into two sections: the 'barn' area, with goafsteads and middlesteads, and the area providing stabling and living accommodation. Supporting the roof over the 'barn' end, are queen-post trusses and collar-beams, with wall-posts and braces. The other area has a king-post truss over the tie-beam. (See Illustration 33.)

These roofs cannot be considered representative of the average farmstead barns, any more than their size.

In complete contrast is the roof of the clay and thatch barn at Washingford Farm, Bergh Apton. This spans an area of only 33 feet by 18 feet, and has 2 tie-beams, each with diagonal struts to the principal rafters. There are no braces.

Another simple construction can be seen on the small cobble and brick barn at Binham, dated 1860. This roof spans an area of 43 feet by 21 feet, and has a low pitch pantile roof, with 4 tie-beams which alternate with collars and struts. (See Illustration 36.)

The steeply-pitched pantile roof of the brick barn at Manor Farm, Kirby Bedon, spans an area of 63 feet by 24 feet. It is very simple in construction: diagonal struts run from the tie-beams to the principal rafters, with braces beneath. (See Illustration 36.)

The 6-stead barn at Church Farm, Alderford, has very rough hand-chiselled timbers. This roof spans an area of 103 feet by 22 feet, and is covered with thatch. The barn has brick walls. (See Illustrations 34 and 37.)

At the barn at Banham Hall, one can see the original oak pegs in the tie beams and rafters. (See Illustration 41.)

THATCH

The best thatching material is Norfolk reed which grows mainly in the Broadland area of Norfolk. Another thatching material – usually seen in the southern regions of the county – is straw.

Reed thatch usually lasts 65–70 years, whereas straw usually lasts 30 years. Reed thatching is about 1 foot thick, whilst straw is about 2½ feet thick: straw thatch is much more loosely packed than reed. The most exposed part of the roof is the ridge: the thatch is therefore thicker here.

Alec Clifton-Taylor writes:

> " . . . reeds, being stiff and brittle, cannot be bent over a ridge, so here a further thickness of thatch was added along the apex of the roof, using sedge, tough grass or straw . . . zigzags and

24. *Tithe barn, Dersingham.*

25. *Barn at Abbey Farm, Barton Bendish.*

scallops are single compared with some of the designs evolved, and here the thatcher really comes into his own. More decoration may be introduced by the rods or spars of split hazel or willow which are often pegged on to the surface of the coping, and in straw thatching along the eaves and barges, too, to give additional security".

The flint and brick barn at Manor Farm, Bacton, has a reed thatch roof with a decorative coping along the ridge. It spans an area of 95 feet by 30 feet.

Reed thatch roofs usually have raised gable ends on the building, to prevent water running through the roof. The barn at Hales Court, which was once thatched, but is now covered with asbestos has crow-step gables. These stand 3 feet above the sheet asbestos: allowing for the thickness of the thatch, the gables would have stood out 1 foot 6 inches. Thatch is also laid to overhang the eaves, since there is no guttering. This was approximately 1½ feet for reed thatch and 3 feet for straw. This was particularly important on dried clay walls, which were not water-repellent like flint or brick.

The thatch is prevented from dropping through the roof by battens, running horizontally and, sometimes, interwoven reeds.

The 3-stead, clay barn at Manor Farm, Tacolneston, has a straw thatch roof with interwoven reeds. The thongs which hold the reeds to the spruce rafters are still visible. The owners of the barn told me that it had taken 10 acres of wheat straw to cover the roof, when they had it rethatched 15 years ago: the pitch of the roof is very steep. The date carved in one of the tie-beams, is 1779.

The man rethatching part of the Paston barn roof last August, told me that to cover an area of 40 feet by 26 feet, as he was then doing, would require approximately 1,000 bundles of reed, and would weigh 3½ tons. (See Illustration 44.) Thatch is a relatively light roofing material and does not require a massive roof structure. It was a material that was readily available in Norfolk. It is a good insulating material, and was once used universally for barn roofs. Rather than replace thatch today, pantiles or asbestos sheeting is used very often: these have the fire-resisting qualities that thatch lacks.

PANTILES

Pantiles are Dutch in origin. The size of pantiles was fixed in 1722, by an Act of Parliament: this was 13½″ × 9½″ × ½″. Local manufacture commenced at the beginning of the 18th century. They were hung over battens by 'nibs'. Pantiles were often made in the same brickyard as the local bricks.

The 5-stead brick barn at Colton, dated 1666, has a pantile roof with crow-step gables at both ends of the building. The roof spans an area of 81 feet by 24 feet.

The barn and granary at Manor Farm, Kirby Bedon, have pantile roofs with 'Dutch' gables. Wrought-iron numbers on the barn wall date it at 1693.

The small chalk and brick barn at Burnham Deepdale has a pantile roof with a plain close verge. The roof spans an area of 48 feet by 20 feet.

Pantile roofs can be seen throughout Norfolk. Many pantile roofs have

26. *Barn at Manor Farm, South Creake.*

27. *Barn at Banham Hall.*

replaced the original thatch. Pantile roofs do not have very good insulating qualities, but they are quite fire-resistant. Unless tiles dropped off, a pantile roof would last longer than a thatched roof. Pantiles have a smaller area of overlap than plain tiles, so the roof is lighter; their design also means that a less steep pitch is needed, and the roof timbers need not be so massive.

SLATE

Slate is not found in Norfolk and must therefore be imported. Any barns with slate roofs, are more likely to have been built by wealthy landowners and farmers.

Slate is a durable, frost- and fire-resisting material and is easy to work. One of the Townshend's barns at East Raynham has a hipped slate roof. The barn is symmetrical, and has brick and cobble walls. It was built in 1870. The lateral sides of the barn are covered with plain, oblong slates, whilst the hips have scalloped-shaped slates. The roof is at a very low pitch: curved red tiles protect the joints over the hip and top ridge.

Coke's Great Barn, at Holkham, has a slate roof. This barn was built in 1790–2 and has brick walls: the roof spans an area of 108 feet by 30 feet. Coke had huge quantities of slate brought to Holkham by sea, from Penryn in Wales. He used them on all the important buildings on his estate.

PLAIN TILES, PINTILES OR PEGTILES

The size of plain clay tiles was fixed in 1477, at 10½″ by 6½″ by ½″.

R. W. Brunskill writes:

> "Plain tiles were nailed, or were hung by means of nibs, moulded into the tile, from light battens. They were laid in regular courses but, in order to protect the joints between tiles, each tile lapped two others, leaving only about four inches exposed and the tiles were usually laid at a pitch of more than 45°.

Plain tiles did not last long as the wooden peg often rotted away and the tile would fall off. They were used mostly in Tudor times. There are few buildings around with plain tiles: I have not yet seen any barns in Norfolk still with plain tile roofs.

FLOORS

In 1782, William Marshall wrote on the subject of Norfolk barns:

> "No farm has less than three threshing floors and these are of unusual dimensions. Twenty-four feet by eighteen is considered as a well-sized floor: twenty by fifteen a small one. . . . Barn floors are of plank, 'whips' (a kind of brick or clay): the last are more prevalent; and although they are considered inferior to the first, they are in better esteem in Norfolk than in most other places, for a Norfolk farmer is aware that what he loses by the handle of his corn threshed on a clay floor, he regains by measure: for the same dust which gives the roughness of the handle in the sample prevents the corn, thus soiled by the day's

28. *Barn at Grey's Farm, Great Witchingham.*

29. *Barn at Washingford Farm, Bergh Apton.*

30. *Grange Barn at Castle Acre Priory.*

beating up, from setting so close in the bushel as that which has been threshed on a clean wooden floor."

The floor area Marshall refers to is the middlestead: the goafsteads would be beaten earth. The wood used for the threshing floor was elm or poplar. It was on these floors that the rhythm of the flails which Ewart Evans describes, could be heard so clearly. I have not yet seen an original elm floor: almost all have been ripped up and concreted for the installation of grain dryers, or implement storage.

Some barns have an extended threshing floor, such as the barn at Ringstead. To cover the floor, a porch would be built over it: these porches usually extend at the front of the barn. Coke's Great Barn at Holkham has an extended porch over each pair of doors.

31. *Tithe barn, Paston*

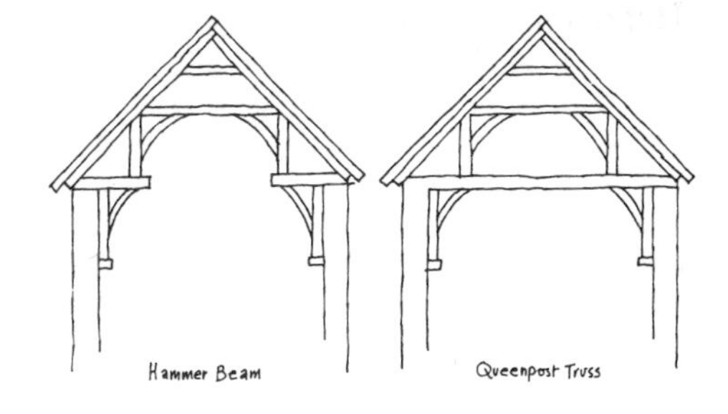

32. *Roof timbers of the tithe barn, Paston.*

33. *Barn at Hales Court*

34. *Barn at Church Farm, Alderford.*

35. *Barn at Banham Hall.*

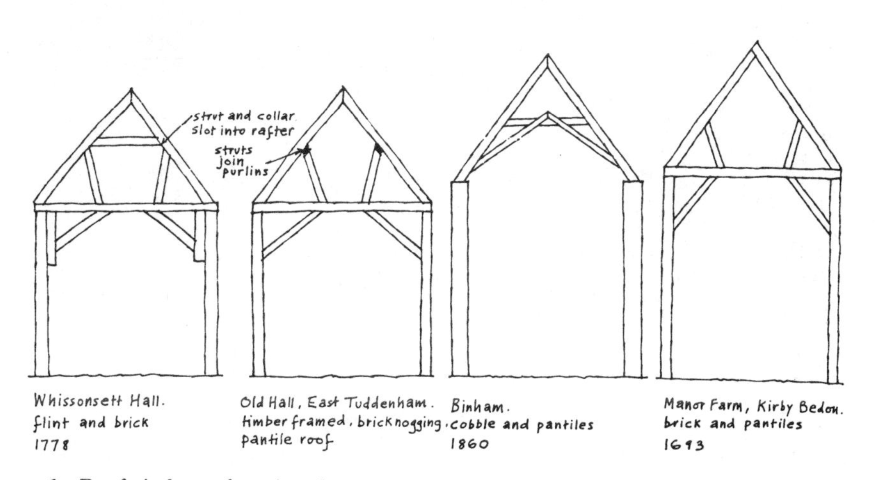

36. *Roof timbers of various barns.*

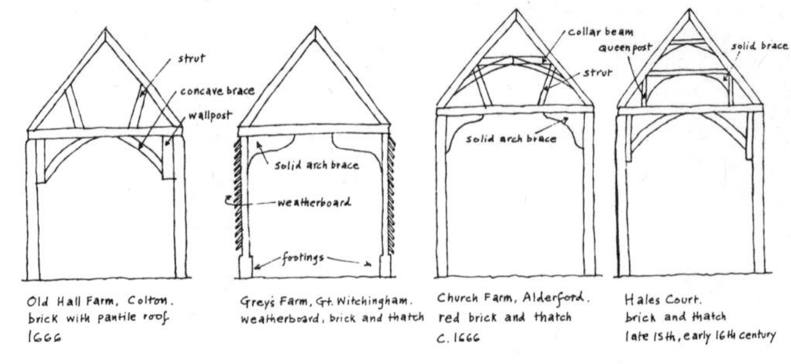

37. *Roof timbers of various barns.*

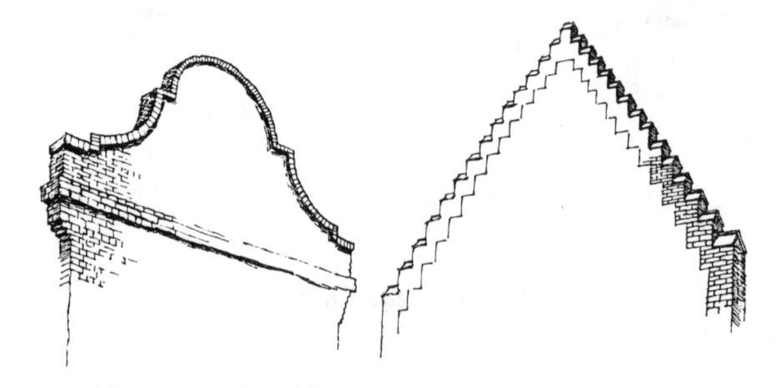

38. *'Dutch' and 'crow-step' gables.*

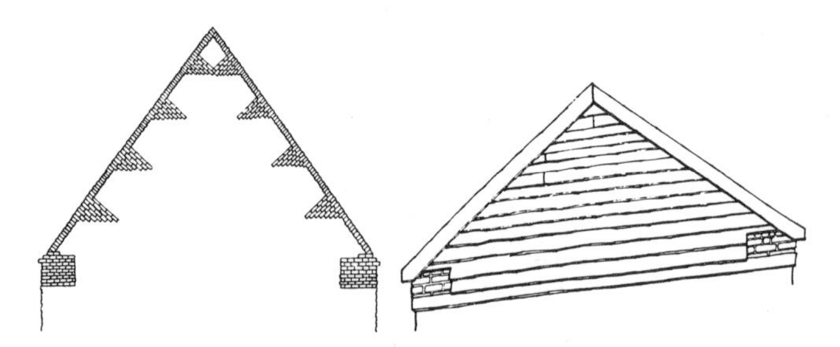

39. *'Tumbling-in' and a weatherboard gable end.*

40. *'Parapet' at the gable and a 'plain close verge'.*